Kindergarten Level

Sudoku Puzzle Book

This book belongs to:

Copyright © 2020 by Andrew Asako
All rights reserved

A humble request

Dear valued customer,

This book was lovingly designed with your full satisfaction in mind.

We, my wife and I, are aspiring writers and self-publishers. Without your help, we would not have a chance to compete against larger corporations with big marketing budgets that we don't have.

Therefore, we make a humble request -if you enjoyed this book- to spare a few minutes to leave us a review on this book's Amazon product page. **Each and every one of your reviews is paramount to us.**

We are forever grateful for your support and we hope we have succeeded in providing you or your loved one with a very special book.

Sincerely,
Andrew and Jenna

Table of Content

How to play sudoku 1
Tips 7
Puzzles and coloring pages 8

HOW TO PLAY SUDOKU

Sudoku is played on a 4x4, 6x6, or 9x9 grid. But, Other sizes exist too!

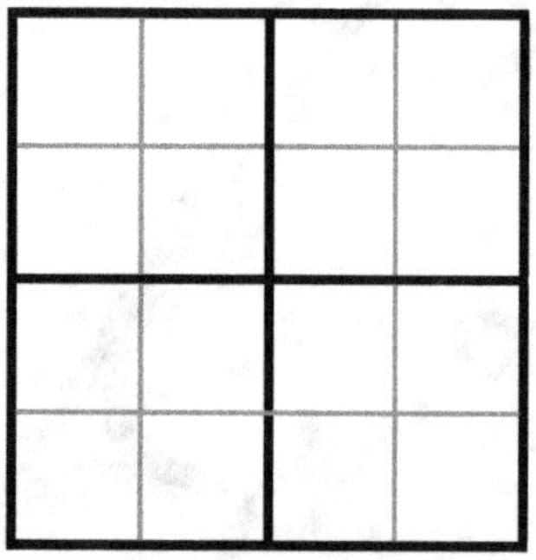

A 4x4 Sudoku grid

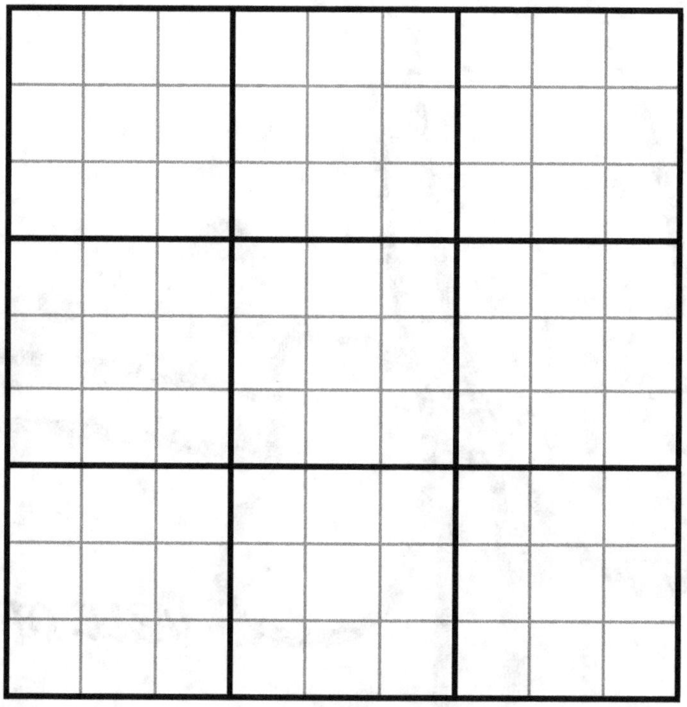

A 9x9 Sudoku grid

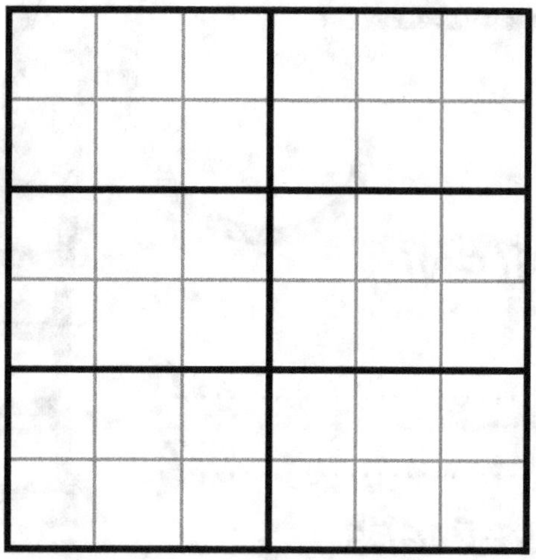

A 6x6 Sudoku grid

For example, a 4x4 grid contains 4 rows and 4 columns, which makes 16 cells. Another thing to pay attention to is what is called "regions", those are the areas inside the bold lines.

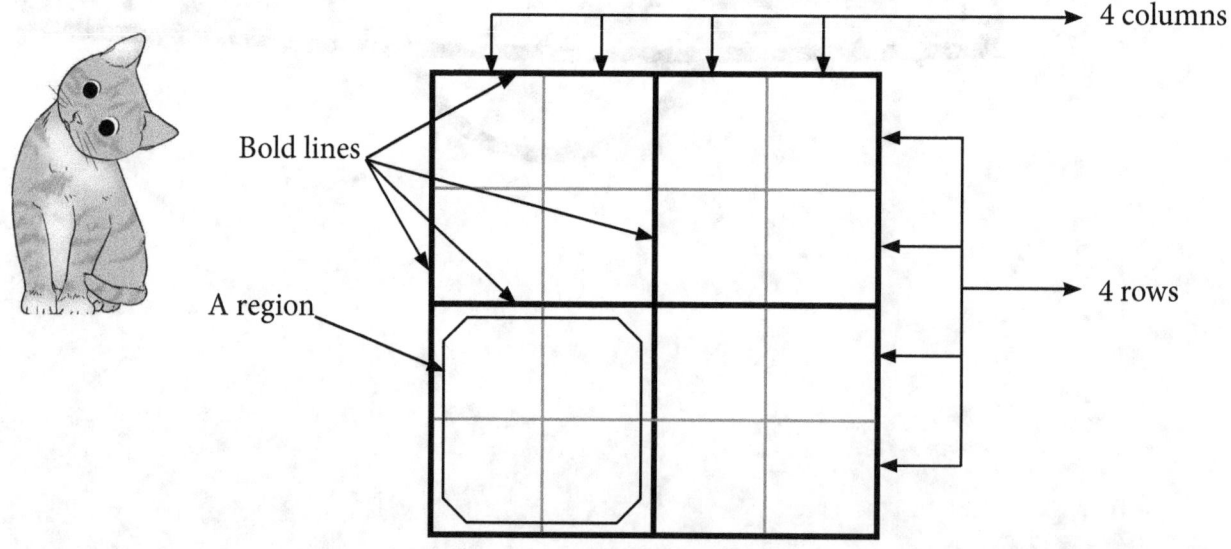

Every sudoku grid comes with a few cells already filled with numbers.

2	3	4	1
	4	3	
3			4

Now that you understand how a Sudoku grid is structured, let's learn how to play!

The goal of a 4x4 sudoku puzzle is to fill each cell with a number from 1 to 4 so that each row, column and region contain all the numbers from 1 to 4 without repeating any numbers in the row, column or region. Sounds complicated? Let's look at an example.

As you can see in the picture below of a 4x4 Sudoku puzzle, the first row is already filled with numbers from 1 to 4. So you'll only have to work on the remaining cells. That means the empty cells at the second, third and fourth rows.

2	3	4	1
	4	3	
3			4

So, lets focus on this cell first

We can see that this cell belongs to second row (Counting from the top) and fourth column (counting from the left).

Now, this is a 4x4 Sudoku puzzle, so the possible values for any cell are the numbers from 1 to 4.

With this in mind, lets check which numbers we can put in.

Step 1: The column rule

The column to which this cell belongs contains two numbers; 1 and 4. So, according to the column rule we can not put those numbers in this cell, otherwise we are repeating them and breaking the rule. 1 and 4 are ruled out.

That leaves us with two possible solutions: 2 and 3.

Step 2: The row rule

Similarly to the column rule, we check the row to which the cell belongs and rule out any existing numbers in that row.

But in this case this row is empty, so it doesn't give us any additional clue. So we still have two possible solutions from step 1.

Step 3: The region rule

The region to which belongs this cell contains the numbers 1 and 4. So the region rule would dictate that those numbers should not be put in the cell.

But we've already ruled out 1 and 4 as possible solutions in step 1.

Therefore, this rule doesn't help us any further in this case.

Step 4: Write down the possible solutions that we've found

Since we couldn't figure out from the three rules in steps 1, 2 and 3 what is the solution, we'll write down the possible solutions that we found: 2 and 3.

To have room for all the possible solutions in the cell, write them down in a smaller size.

Now repeat steps 1 through 4 for all the empty cells.

(Note: as you might have notice, you can change the order of steps 1, 2 and 3. It doesn't make any difference)

You should get this ⟶

Now, notice those four cells

They have only one possible solution!

Congratulations! You have solved your first cells.

Lets write down those solutions in normal size

Page 4

Now you can go through the three rules again (column, row and region rule) but this time try to do it in your head without writing and see if you can solve other cells.

For example, using the column rule we can solve this cell

From the two possible solutions (1 and 4) we can rule out 1 because it already exist in the column to which the cell belongs. So the answer is 4.

Using the region rule, we'll solve this cell

2 and 3 are the possible solutions but 2 already exists in the region to which the cell belongs, so 2 is ruled out. That leaves us with 3 as the solution.

Keep using the tree rules that you've learned and you'll solve the entire puzzle! If you did, you can take a break and color the drawing in the next page!

COLORING PAGE

WHAT YOU SHOULD KNOW BEFORE YOU START

Don't guess

Sudoku is a game of logic and reasoning, so you shouldn't have to guess. If you don't know what number to put in a certain cell, keep scanning the other areas of the grid until you see an opportunity to place a number. But don't try to force anything. Sudoku rewards patience and thinking, avoid relying blind luck or guessing.

What to do when you are stuck

You can always go back and revisit the example in the previous pages if you forgot how to solve Sudoku puzzles.

PUZZLES AND COLORING PAGES

Puzzle #1

		3	4
	4		1
	2		3
		1	2

Puzzle #2

	4		3
3			1
4			2
2		1	

Page 9

COLORING PAGE

Puzzle #3

	3	4	
2			1
4	2	1	3

Puzzle #4

		2	
2	4	1	
	3	4	2
	2		

Page 11

COLORING PAGE

Puzzle #5

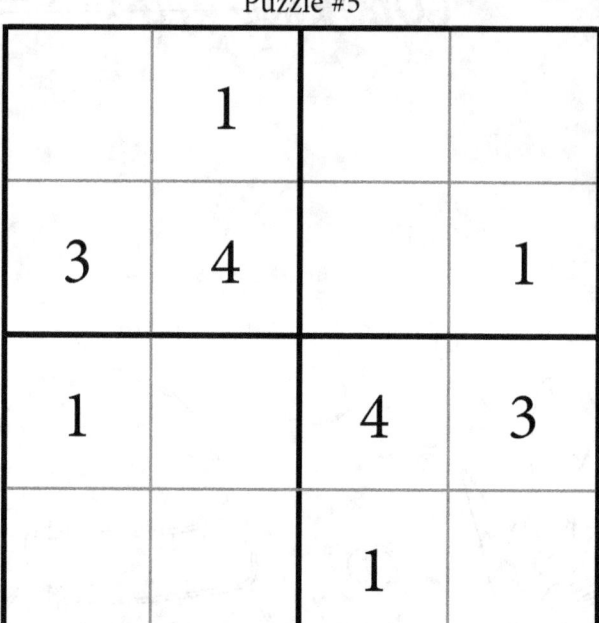

Puzzle #6

COLORING PAGE

Puzzle #7

			1
	2	3	4
3	4	1	
2			

Puzzle #8

1			3
2		1	
	1		2
3			1

COLORING PAGE

Puzzle #9

	4	3	
2			1
3			4
	1	2	

Puzzle #10

	3	4	
1			2
	2	1	
4			3

Page 17

COLORING PAGE

Puzzle #11

	4	2	
1			3
2	3	1	4

Puzzle #12

	4		2
	1		
3		4	

COLORING PAGE

Puzzle #13

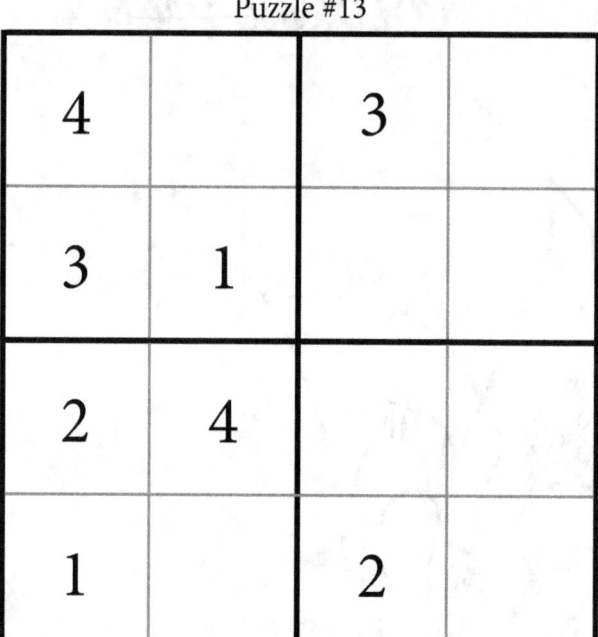

Puzzle #14

Page.21

COLORING PAGE

Puzzle #15

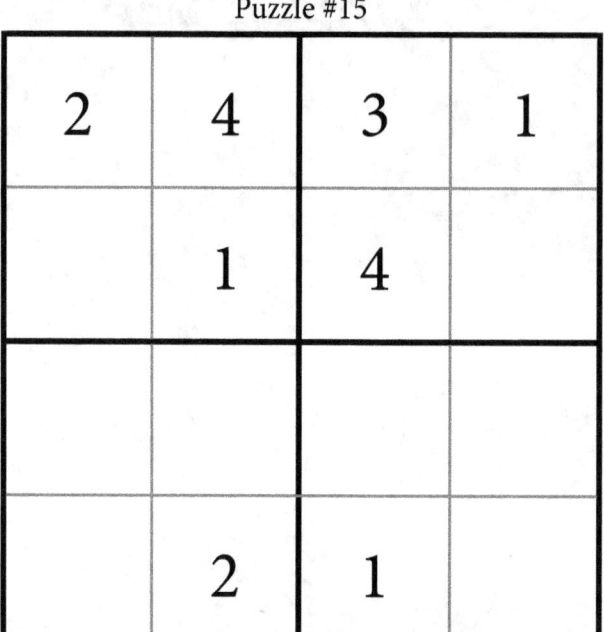

Puzzle #16

COLORING PAGE

Puzzle #17

1			2
2	3	1	4
3			1

Puzzle #18

		1	
2	1		3
1		3	4
	3		

COLORING PAGE

Puzzle #19

3	4	2	1
2	1	3	4

Puzzle #20

	4		
	3	2	
	2	1	3

Page 27

COLORING PAGE

Puzzle #21

2	4		3
3			
			1
1		2	4

Puzzle #22

4			1
	3		4
	1		2
2			3

COLORING PAGE

Puzzle #23

	4	2	
1	2		
4	3		
	1	3	

Puzzle #24

	4		3
3	2		
		1	4
4		3	

Page 31

COLORING PAGE

Puzzle #25

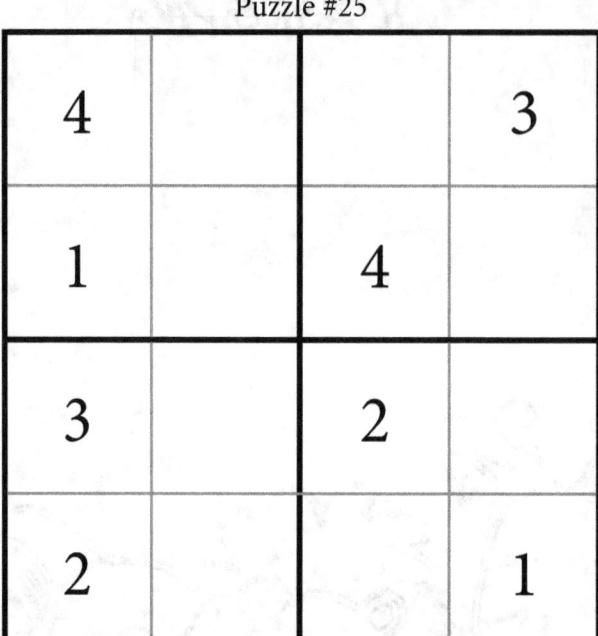

Puzzle #26

COLORING PAGE

Puzzle #27

	1	4	
2			3
4	2	3	1

Puzzle #28

4	1		3
3		1	4
1	4	3	

COLORING PAGE

Puzzle #29

1			3
4		1	
3		2	
2			4

Puzzle #30

2		4	1
			3
1			
3	4		2

Page 37

COLORING PAGE

Puzzle #31

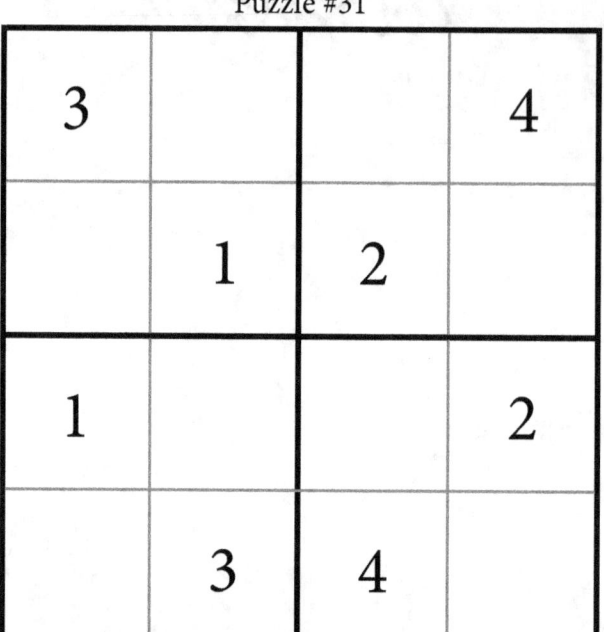

Puzzle #32

COLORING PAGE

Puzzle #33

2			
4		2	
1	4	3	2
	2		4

Puzzle #34

	1	4	
3			2
4			1
	2	3	

COLORING PAGE

Puzzle #35

3			1
	4	2	
4	1	3	2

Puzzle #36

			2
2		1	4
3		2	1
			3

Page 43

COLORING PAGE

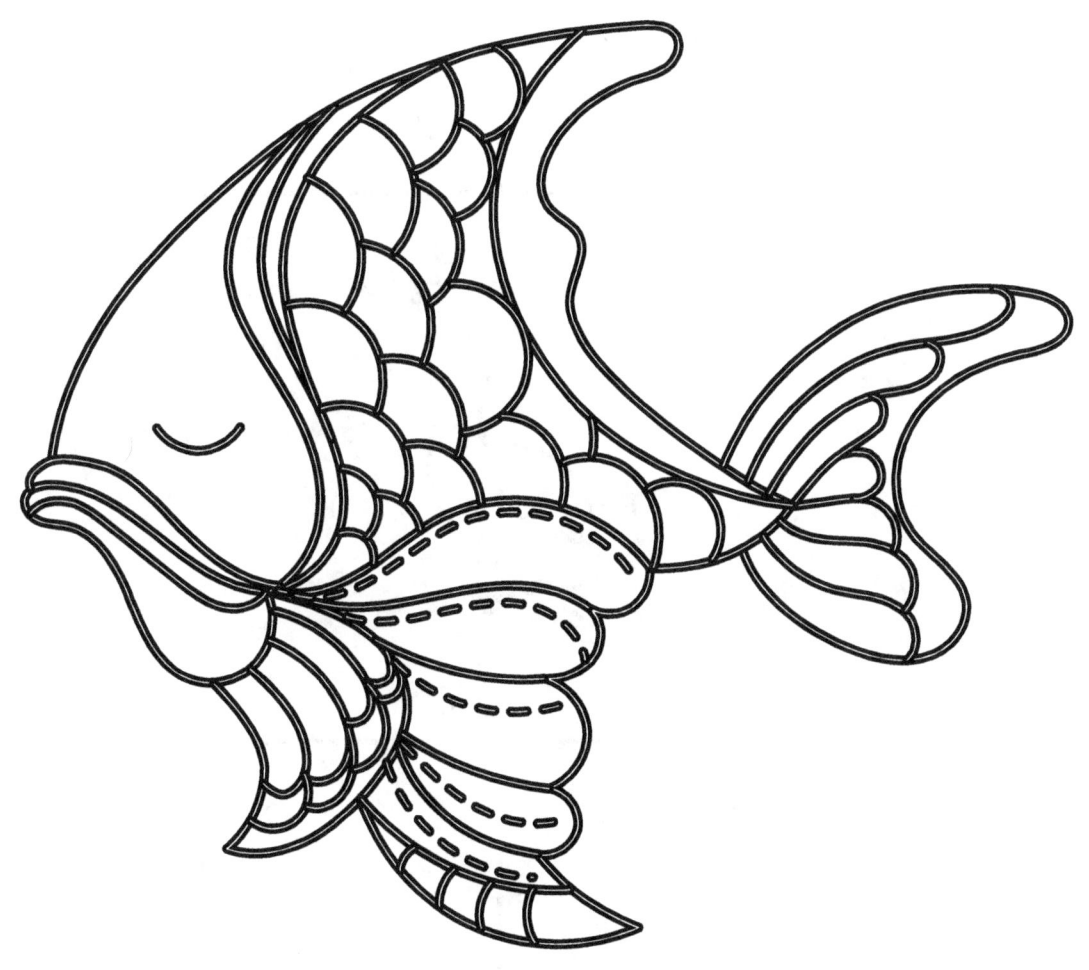

Puzzle #37

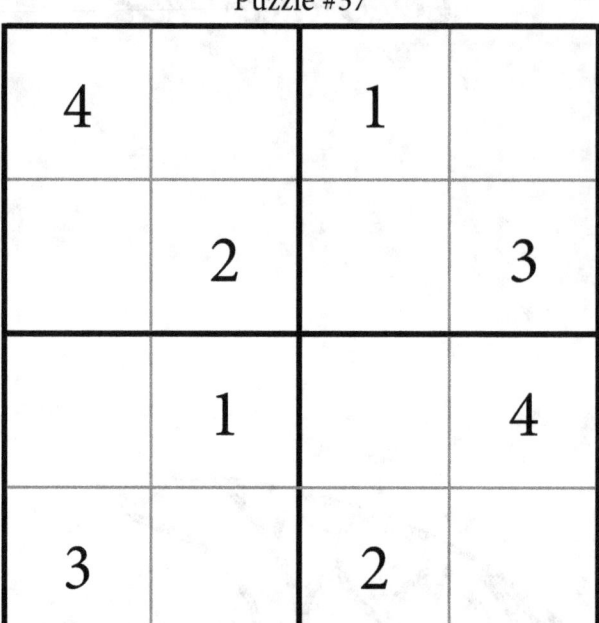

Puzzle #38

Page 45

COLORING PAGE

Puzzle #39

3		1	
	1		4
	2		3
4		2	

Puzzle #40

	2	4	
4	3	1	2
3			1

COLORING PAGE

Puzzle #41

		1	2
	1	3	
	4	2	
		4	1

Puzzle #42

1		3	
2			1
3			4
4		2	

Page 49

COLORING PAGE

Puzzle #43

	4	2	1
			4
1			
4	3	1	

Puzzle #44

		3	
3		4	2
2	4		3
	3		

COLORING PAGE

Puzzle #45

	1	4	2
			1
			3
	3	1	4

Puzzle #46

4	3	1	
	1		
	4		
1	2	4	

Page 53

COLORING PAGE

Puzzle #47

	4	2	3
		4	
		3	
	3	1	4

Puzzle #48

		4	2
	4	3	
	3	2	
4	2		

COLORING PAGE

Puzzle #49

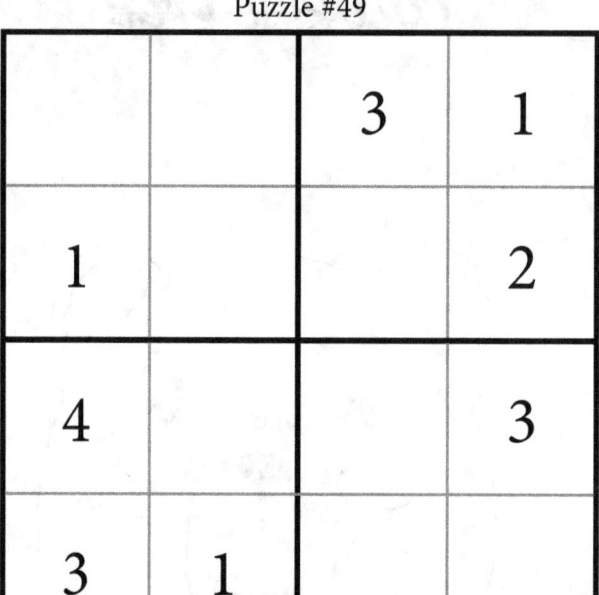

Puzzle #50

COLORING PAGE

www.ingramcontent.com/pod-product-compliance
Lightning Source LLC
Chambersburg PA
CBHW060440220526

45465CB00008B/3218